Praise for Rebecca
Place in the Spiral

"I am here. I once was here. I will return here. The here always remains," states author Rebecca Beardsall in her part memoir/part photo album *My Place in the Spiral*. In the author's insightful and intriguing journey to research and spiral back to two women ahead of their times—her grandmother Ruth and great-grandmother Mary—Beardsall forges for us a path of understanding. Comparing faces, mannerisms, conversations, houses, educations, beliefs, superstitions, and dreams, she leads us from her New Zealand and Western Washington homes back to her Pennsylvania German heritage.

We, too, are in these pages, detectives uncovering clues to better understand, perhaps, our own identities. Mennonite upbringing re-stitched with feminism and literary theory? The future superimposed with sepia-toned photographs? Yes. In *My Place in the Spiral*, "the past...[serves as] a vision of [the] future....[t]he gyre of memory... looping back again." In these pages, Rebecca Beardsall gives us the people we love alongside the ancestors we may never have met. In doing so, she encourages us to rediscover in them our present and future lives.

–Marjorie Maddox, www.marjoriemaddox.com, author of the prose collection *What She Was Saying*

Through Beardsall's use of photographs and narrative captions, it's as if we have all been invited to an intimate family slideshow. *My Place in the Spiral* begins with a look at time, at memory, and at our place in all of it and ends with the satisfaction that Beardsall has found herself, her nana, and her great-grandmother connected in the those spiraled lines that are always retreating and returning all at once.

Rebecca Beardsall's quest to find out how and why she has always connected with her nana and great-grandmother is a literary journey through photographs and travel. With each turn of the page, we see what she sees, the closing of the distant connection between two women she had always wanted to meet but couldn't and the warmth that grows inside of her with every discovery that she is more like them than she could have imagined.

—Kase Johnstun, http://kasejohnstun.com/, author of *Beyond the Grip of Craniosynostosis*

My Place in the Spiral is ostensibly about her search to find out something about her grandmother. And it is that, and that simple story is interesting enough in its own right. But it's also about more far more than that. Through photographs and footnotes, the book asks us to suspend ourselves in multiple moments in the same moment, to see one body in multiple bodies (or is it multiple bodies in one body) and, in doing so, to confront any number of quietly sublime truths about our complicated relationship to time. At various points, the book reminded me of Mitchell, Vonnegut, and Dickens; yet the book goes beyond those now dusty meditations on time to propose yet a new relationship to time.

Beardsall uses family history to bend time back on itself so effortlessly. The story runs through your hand a bit like sunlight or cold river water or time itself: beautiful, important, impossible to capture or contain, let alone describe.

Readers will find themselves in *My Place in the Spiral*, I think, because we have all looked into a mirror and watched an unexpected ancestor peer back out. Time does not progress. It swirls. It eddies. It flows faster. Sometimes it stops, even doubles back on itself.

—Nathan R. Elliott, Ph.D

Memoir takes a visual, time-traveling and always intriguing interpretation as Rebecca Beardsall's book crisscrosses family, history and destiny in sometimes startling discoveries that inspire further exploration of the mysteries in one's own memories.

—Sarah Eden Wallace, multimedia journalist, Falling Star Studio

My Place in the Spiral

REBECCA BEARDSALL

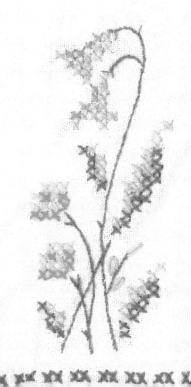

For Marilyn and Pat

Introduction

It is hard to walk down the street without walking into a photograph in progress. For the most part the posing is obvious, and I am able to maneuver around them. But sometimes they stop abruptly in the middle of the footpath causing all of us, who are trying to get to and from work, to stop and wait. Sometimes we will get the universal hand wave to let us know it is safe to proceed, or we wait for the camera to lower before we walk on. I often wonder how many photo albums around the globe contain my likeness as I go about my daily life in Auckland.

It is a beautiful city and a beautiful country. I can understand their desire to capture it. For many, it is a long journey to arrive in the South Pacific, with New Zealand/Aotearoa being the furthest south in Oceania. The naming often contributed to the legendary figure Kupe when after leaving Hawaiki, he spied in the horizon a long cloud in the sky and the land below – Aotearoa, Land of the Long White Cloud. I live on the North Island, also known as Te Ika-a-Māui, meaning the fish of Maui in Māori. I believe, on some level, this is what people are trying to capture in their photos of New Zealand/Aotearoa – not just the beauty of the place but the rich Oceania history. It sounds cliché, but there is something magical about this place. This land. Kupe felt it on arrival, and every foot that has stepped on these shores since feels it. So, I'm thankful for the people who do take photos of the place where I live because I haven't taken the time to stop and see my city through a camera lens. Oh, I have some pictures from when my parents came to visit, but mainly my photos replicate scenes outside the city when we venture to places like Taupo or Paihia.

The bright blue waters of Waitematā Harbour highlight the City of Sails. Camera lens focus on the CBD as the sunlight glints off the buildings. I am on the ferry heading to Devonport, and I look back at the city; watch the people pose on the boat with the Auckland skyline behind them. The Sky Tower stands tall like a proud kauri tree. It is a focal point of the skyline, and people continue to aim their cameras at it, snapping photos which will end up in a vacation album somewhere. They will look fondly at it and remember their time spent in New Zealand/Aotearoa. The Sky Tower isn't just a focal point in a photo for me, but it indicates home. It is an easy marker, like a red pin on a map, to show us home. Our apartment is about 400 meters away from the Sky Tower. I use the Tower as a way to direct people to our apartment. If you know where the Tower is, you know how to get to our home.

• • •

It has been years since we lived in that apartment, and many years since we lived in New Zealand/Aotearoa; however, even today, once again heading to Devonport, looking back from the Fullers ferry at the city I call home, I feel a sense of groundedness knowing that the Tower points to a place I lived. It exists. It is part of me. I may have moved on, but this place, this city, this country is still home. From the Sky Tower, I can follow the outline of the buildings until I see where our apartment stands, and I feel content. The mapping of the place confirms that I know this city. I have intimate knowledge of this place that the people who just arrived and snapped photos of it can never know. They can't call this place home. They might hear me with my reverted American accent and think I'm a tourist too, but my vowels once had the Kiwi twang, and I once walked Queen Street to go home.

I am here. I once was here. I will return here. The here always remains.

• • •

Te tōrino haere whakamua, whakamuri in Māori means: At the same time as the spiral is going forward, it is returning. The spiral imagery is integral to Pacific cultures; it is in tattoos, artwork, myth, and story. It is the fern frond depicting new life. It is the gyre in the ocean representing the currents that connect us. It is also the way Māori view time.

It wasn't until I read Epeli Hau'ofa's book *We Are the Ocean*, where he discusses the circular nature of life, that I realized my way of looking at the world wasn't unusual; instead, it was just different from my culture's understanding of time. Instead of placing so much value on the future, Oceanic cultures emphasize the past. Ever since I was a child, I was intrigued by the past and the way it existed in the memories and stories of my parents, aunts, uncles, and grandparents. Often my insistent need to understand what came before was brushed off as nostalgia or romanticism. American culture and society looked forward – always forward – not back. When I tried to connect with the past, I heard - why care about the past when the excitement of the new was in the future? When I read that Oceania cultures valued the past to understand the present, I shouted "yes" loud enough to startle the cat sleeping beside me. Here I found the theory, terminology, and foundation for an understanding of time that had existed inside of me. It wasn't Western linear time, but circular time and later the spiral time that would bring me into a new understanding of myself and the world around me.

In the chapter "Pasts to Remember" Hau'ofa states, "That the past is ahead, in front of us, is a conception of time that helps us retain our memories and be aware of its presence" (67). He stresses the need for people to investigate and understand their past so that they can not only situate themselves in the present but can move into the future. For most Western cultures, time is viewed as linear in nature – pushing the past behind and looking forward to the future. Oceania cultures view time in a circular nature with the past leading the way and always in the forefront. Hau'ofa notes:

> From this perspective we can see the notion of time as being circular. This notion fits perfectly with the regular cycles of natural occurrences that punctuated important activities, particularly those of a productive and ritual/religious nature that consumed most of the expended human energy in the Oceanian past and still do in many parts of our region today. (67)

The circular understanding of time helps individuals stay connected to their past, their history. According to Hau'ofa, "What is ahead of us cannot be forgotten so readily or ignored, for it is in front of our minds' eyes, always reminding us of its presence. Since the past is alive in us, the dead are alive – we are our history" (67). Not only is it history, but it is also memory. To remember people – the past, the stories, the myths – is a place to house the present and ultimately the future.

Māori spiral time takes this understanding of time a bit further in the sense that instead of time moving in a continuous circle, it spirals in and out simultaneously. Witi Ihimaera, a Māori author who utilizes the spiral understanding in his writing, expresses the concept of time like this, "The double spiral . . . allows you then to go back into history and then come out again. Back from personal into political and then come out again" (qtd. in DeLoughrey 162). The spiral creates the space for time – past, present, future – to meld together into moments. Our ancestors directing, guiding, defining as if they were here in the now and in the future.

I believe we all live within the spiral of time, even if we don't recognize it or understand it. When people say to me, "That was in the past, it isn't my problem;" "I'm not responsible for something that happened a hundred years ago;" "Why should I apologize for something my ancestors did? I didn't do it," I try to talk to them about the spiral. We need to understand and acknowledge our past; that doesn't mean we have to agree with it, but until we acknowledge it, we can't move beyond it. The spiral will continue to return it. Serene Jones, in her interview on the podcast *On Being with Krista Tippett*, contemplates moments of grief, trauma, and grace:

> Well, one of the things that I learned from life, but it came to the fore in writing the book on trauma, is that not just individuals, but whole communities undergo trauma, and that one of the characteristics of trauma is the deep human desire to repress it and to not deal with the story of the harms that have happened. But the truth of the matter is, with individual and collective trauma, is that the harm haunts you, haunts your dreams as an individual, haunts your collective unconscious as a society, until you tell the story, till you face the truth of the horrors that have happened. And I think what is happening in our nation today is, all of the harms of the past have come up to claim us, all at once, and they're not gonna let us go until we take the stride of reckoning with them.

Jones is speaking about the spiral of time, even if she isn't using that terminology. The past is always in front of us, and the Māori understanding of the spiral brings that into clarity. Elizabeth M. DeLoughrey, in her book *Routes and Roots: Navigating Caribbean and Pacific Island Literatures*, investigates the way the spiral exists within Māori culture and narratives. DeLoughrey, in the chapter "Dead Reckoning," states, "The spiral gestures to the past while moving into the future, positioning historical events in the present so that time becomes coeval or simultaneous" (188). We are our past, our history. Here. Now.

• • •

Even though I valued the past and always enjoyed listening to the stories from my family, there was a portion of my history that was story-less. It wasn't until 2015 that I realized I knew nothing about my maternal grandmother and her parents. Ruth, or Nana, was a photo in our living room. For most of my childhood, her photo was placed near the television, the focal point for the family when we were gathered together in the room. Later the picture of this woman with wavy hair, glasses, and an infectious smile hung next to the old school clock. A wooden clock with a glass door that used to manage the time of the teachers and students at Collegeville-Trappe High School. Whenever I looked up to check the time, there she was. I knew the likeness imprinted on glossy paper was Ruth. My sister's middle name. The name of my mother's mother.

She remained there, on the wall by the clock, until I realized there was a hole in my past and my understanding of myself. I needed to enter my own spiral of time. The quest and the journey connecting me to Ruth, and later Mary, transported me to places and emotions I never expected. My life is a continuation of theirs. I am a moment in their spiral and they are a moment in my spiral. Our lives are together simultaneous. Echoes reach back and forth to each other, confirming our connection.

I am here. I once was here.

Rebecca Beardsall
2020
Bellingham, Washington

"I remember from childhood that, from the point of view of a child, a mother is a fixed entity, a monolith, not a changing evolving human organism who is qualitatively similar, in many ways to a young person."

- Sarah Manguso, *Ongoingness*

"I was traveling backward in time toward myself at the same time I journeyed forward, like the new star astronomers found that traveled in two directions at once."

- Linda Hogan, *Solar Storms*

Kia ū ki tōu kāwai tūpuna, kia mātauria ai, i ahu mai koe i hea, e anga ana koe ko hea.

Trace out your ancestral stem, so that it may be known where you come from and in which direction you are going.

- Māori proverb

Thai Bowl at R. Thomas Deluxe Grill, Atlanta,
The meal that required a picture, thus leading to a discovery never imagined.

Rebecca (left) and Tracey (right), November 2014; R. Thomas Deluxe Grill, Atlanta.
The photo that started it all[1]

---

[1] I posted this photograph, and Aunt Pat replied immediately, "Rebecca, I looked at your picture and saw Ruth looking at me." Ruth, my nana, was someone I've never met. She died three years before I was born. I laughed at my Aunt's comment and said it must be the glasses.

Red dress that started the investigation, July 2015[2]

---

[2] Eight months later, I was proud of myself for being able to fit into a dress I ordered online around the time I went to Atlanta. When the dress arrived, it fit, but it was snug. I started with a focused effort to lose weight, for health reasons, at the beginning of July. It was the 21st of July and I thought, *I think I can fit into that red dress now*. I held my breath as I slipped the dress over my head. I paused in front of the mirror to see that it draped my body perfectly and I no longer looked like a squished sausage. I could finally see some real change in my body, so I put on some red lipstick to go with my new-ish red dress. I wanted to capture the moment, so I took a photo of myself before heading to work. I posted the photo to my social media accounts. My mom, this time, immediately responded with, "You look like your Nana in this picture."

This was the first time I started to wonder about their comments. My mom has often compared me to her mother, but never about my appearance, more so on my intellect. She said my love of going to school and passion for learning matched Nana's outlook on life. However, I followed the humanities track while Nana enjoyed mathematics and sciences. I attribute my uncanny comprehension of calculus to my nana. It was one of the classes in high school

Ruth (left) and Rebecca (right)[3]

that I just understood without studying. My classmates would ask me how I knew the answer and my reply to everyone's bafflement, including mine, was, "I don't know. I just do." Otherwise I'm horrible at math; therefore, I can only attribute my excelling in calculus to the mystery of genes.

When I graduated from DeSales University with my BA in English, my mom gave me her mother's desk. It was the desk Ruth received on her high school graduation from her parents. In fact, I'm sitting in Ruth's desk chair as I type this piece. So it was not out of the ordinary to hear my name and Ruth's at the same time, but this was only the second time that I was told that I looked like her. I didn't see it. The woman I remembered from photos in family albums and sitting in a frame by the old clock didn't look anything like me. I didn't know what my aunt and my mother were talking about, but the fact that they both said it made me pause.

[3] I looked at Ruth and still did not see myself. Yes, the glasses, but that was about all that I could find that made us look similar. I pulled out a photo with Ruth in a similar pose to my own in the red dress. I inhaled the shock when I placed the photos side-by-side. She was staring back at me through my own eyes. I gathered more photos of Ruth and placed them next to my own. I could not believe it. How was it possible that I had gone this long without seeing it? Why now?

Ruth (left) and Rebecca (right)[4]

It gave me a place and a space to see – literally – see how I was connected to this family. Family friends often said that I did not look like my siblings. My brother looked like my pop-pop and even followed many of his mannerisms. My sister looked more like my Aunt Pat – so closely, in fact, that when we visited Aunt Pat when my sister was in high school, people often thought my sister and Aunt Pat were mother/daughter or sisters. I was the outsider, not really linked to anyone.

[4] My understanding of myself started to shift when I realized I was connected to this woman named Ruth. This woman whom I heard so much about – no scratch that – it is not true. I only had bits and pieces about her. I remember her picture being on our living room wall when I was growing up. I knew that she was a seamstress and made the quilt on my parents' bed when she was, something crazy, like eight years old (see note 1). I knew

Ruth – beautiful and studious. High School senior photo.[5]

she went to college, and I knew she was a teacher. I knew that she had once served two large platters of corn to the family, and called it dinner, on one of the occasions when my father dined with the Gottshalls. I had her desk and her chair. I also had her mother's library table sitting in my own library where it serves as a focal point housing the books I published and books from New Zealand. Recently, I added an old Remington typewriter I received on my 39th birthday – my first birthday celebrated in Bellingham, WA. That was about all I knew about this woman who shared my smile.

[5] It occurred to me – why did a girl who grew up on a farm decide to go to college in the 1930s? This one question led me down a path of more questions. I picked up my phone and started texting my mom and my aunt. I peppered them with questions about their mother, their grandmother and grandfather. I wanted to know everything. I needed to know more than anything what made Ruth tick. If I was so like her and even looked like her, maybe, just maybe I could learn more about myself as I discovered her. I also wanted to learn more about Mary, the mother who shaped (maybe), influenced (possibly), and raised (indeed) my nana. Sadly, I was unable to get a lot of answers out of Mom and Aunt Pat. They could give me my great-grandparents' names, but nothing else.

They remembered my great-grandmother tutoring, so they assumed she taught during World War I. They said she was a seamstress for a high-end store in Norristown. My great-grandfather, E. Leroy, died young from Leukemia. In the messages that were

Ruth in the garden, June 16, 1935, wearing a dress I can only assume she made.[6]

bouncing back and forth from Washington to Pennsylvania, I found out Ruth had a sister, Marion. I don't remember knowing this but when my mom texted that we went to Aunt Marion's on the way home from Chincoteague one year, I asked if it was a house with a circular driveway that had large white planters out front. Her reply was yes – and that is all I remember about Great-Aunt Marion. I have a vague memory of iced tea at this house as well, but I don't know if it is a real memory or something transposed from another moment in time. The tea had a flavor like jasmine or anise… something a child would find strange. I remember not liking it, but the taste seems so familiar now that I'm sure if I was offered a glass today I would find it delightful.

I found out that Ruth went to Ursinus College and Marion went to West Chester.

Wait. They both went to college. Two girls in the mid-1930's went to college? Why? What was their home life like that would make college an option when college wasn't even a real consideration or necessarily a track for me and my siblings in the 1980s and 1990s?

[6] I grew up in Pennsylvania just like Ruth; however, attending college was not the norm. It was the 1990s – 56 years after Ruth graduated from high school. Yes, going to college was to be the trend by the time my classmates and I graduated, but it wasn't expected. Basically, in the small Mennonite circle where I was brought up, the main goal of a young Mennonite girl was to find a man, get married and have babies – maybe lead a Bible study – well, a women's

Following Mom[7]

Bible study, because there was no way a woman, like myself, could teach men. It all felt a little bit like Victorian England. I had visions of white gloves, Sunday hats, and walks in the garden.

[7] For the most part, my young life revolved around recreating my mother's life within my own context. I had a half-size washline installed on the patio so I could hang up my doll clothing, just like Mom. My Fisher-Price grocery cart got loaded into the trunk of our car on shopping days so I could walk around the store, just like Mom. I cared for my dolls with great seriousness. I was their mother, teacher, librarian, just like Mom. I saw my life as a parallel to my mother's. The church also defined these roles for me. Every Wednesday night I went to Pioneer Girls, where we were taught crafts, sewing, and cooking. They were making us into homemakers living in community with other women of the faith.

When I was sixteen, I was asked to pick out my cutlery pattern and dinnerware. I dutifully selected what my teenage self thought was interesting, and I received dishes and cutlery for the next few Christmases. White dinnerware with green edges and ivy, because when I was sixteen, I dreamed of ivy covered buildings. I finally released myself from those dishes in 2011– well, most of them. I still have all the serving ware – when I gave them to my mom. The cutlery is used every day in my home. I often pause to study the pattern, which seems counter to the person I've become, and wonder about the tradition of setting up a hope chest for a wife in training.

At the age of nineteen, I was living in Scotland and on the hunt for a husband, because I

6/16/35

Ruth in the garden, June 16, 1935. High School Graduation[8]

thought that was all that was ahead of me. Someone said to me, "But you are so young, why are you thinking about marriage?"

My answer was, "What else am I supposed to do?"

I often think about this nineteen-year-old girl thinking life revolved around marriage and babies. If she only knew then that she would later become a feminist – gasp, a curse word in the church where she was raised – and she would never have children of her own. Maybe then she could have soaked up the experience of living abroad without the nagging fear that time was running out for her.

Recently, I realized out of the three kids I spent the most time shadowing my mom. Maybe it was because I was the youngest or maybe because I just liked to watch and learn. I was the only child that enjoyed helping in the kitchen: My sister avoided it, and my brother only came in if there were samples. I followed my mom around as she worked in the garden and planted flowers. Partly because it meant I got to control the hose and fill the freshly dug hole with water. In all this, I was also the only child that had a mother who worked. My siblings had a stay-at-home mom, and I did, too, for part of my childhood. So maybe my childhood mirrored the matriarchal line more than I comprehended as a young adult.

[8] So how did Ruth in 1935 know she was going to college and not have the same mindset I did at that age? I was raised by one of her daughters, so why would I think my mom was raised any differently? I reached out to Mom to try to understand how a Mennonite in 1935 thought higher education was the right path for a woman – and that is where my assumption

Nana's brick-like Bible. Did it hold any answers?[9]

was corrected. Oh no, Nana was…Lutheran, not Mennonite. So once again, something I assumed to be true was shattered. The Mennonite heritage came from Pop-pop's side. Our family narrative was so deeply rooted in the understanding that our ancestors were instrumental in bringing the Mennonite faith to Pennsylvania. Rev. Jacob Gottschalk was the first Bishop of the Mennonite Church of America, and a spreading collection of lines connected my life to his. Seeing our family name in stained glass windows and on plaques throughout Pennsylvania Mennonite churches only cemented this lineage in my mind. I never once thought, growing up, that we were anything but Mennonite. I don't know why, because my father and his family were Lutheran; my uncle was a Lutheran minister. Yet, I lingered in the Mennonite ethos. Maybe it was because it was the church I was raised in, and the Lutheran religion and experience seemed foreign to me. To learn now that Nana was only connected to the Mennonite faith through marriage startled me. It shouldn't have, but it did all the same. I started to see how our lives and the principles that guided them could have been different. She joined the Mennonite church as an adult, or at least I assume she joined, and all I knew was the Mennonite framework and its teaching on gender roles.

I received an email from my aunt around this time in response to some of my questions. She wrote: "I think the one reason that Mom, Ruth, went to Ursinus was location. She was the only day student in her class. This was permitted because the family farm butted up to the campus. Ruth majored in Biology, Math and German at Ursinus, but could not get a high school teaching job when she graduated in 1939, because she was female. She worked at Curtis Publishing in Philadelphia as a statistician. Mom was a wonderful seamstress, had perfect penmanship, and played piano beautifully."

[9] During my quest to learn more about Nana, Mom sent me Ruth's Bible. The thickest, most

brick-like Bible I've seen (and I've seen many – I used to work at a Christian bookstore where I sold many Bibles in all shapes and sizes). I tried to imagine Nana carrying this beast of a Bible around. It was almost three inches thick. Opening the first page, I noticed it was from The National Bible Press in Philadelphia (shopping local, I see) from 1943. The type – bourgeois. I grinned – seemed like an interesting choice of font type (and name) for a Bible. I started flipping the pages looking for a clue. On the pages of 2 Kings 24-25 I found a remnant of a leaf or something stuck down deep within the spine.

Although the church confounds me with its hypocrisy and patriarchy, it is still where I'm rooted. It is my history, my story. It exists in my spiral. Nana's Bible has a couple of pages repaired with tape. The largest section of tape in the Bible is on the tab section of 1 Samuel Chapters 2-3. (1 Samuel is the book of the Bible that follows Ruth – naming the order of books of the Bible is still engrained in my mind thanks to the songs from Sunday school). 1 Samuel 2: 1-10, Hannah's Song, was once spoken into my life from elders when I lived in Scotland. Yet, this is not what stands out to me in the place where Nana's tape marks the page. No, not Hannah's song, but Chapter 3, when the Lord calls Samuel. This has to be one of my all-time favorite Bible stories. While most kids wanted to hear the story of Noah and the Ark, Jonah and the Whale, David and Goliath, or the Christmas Story (birth of Christ), I returned to and requested the calling of Samuel. The story set-up tells the reader that this is a time when messages and visions from God were rare. (A good clue it is all about to change.) In the night, Samuel hears his name called and runs to Eli, his elder and mentor, saying "Here I am." But Eli tells the boy he didn't call him. This happens two more times during the night. The third time, Samuel wakes up Eli stating, "Here I am; you called me."

Eli realizes Samuel is hearing the voice of God. Eli tells Samuel what to do the next time he hears his name called. When God calls, "Samuel," this time Samuel answers: "Speak, for your servant is listening." (As he was instructed by Eli). This story still makes me smile. I love the multiple times he runs to Eli announcing his presence: "I am here." But all the time, that wasn't where he needed to be. It was a different here. He needed to open himself up to the new here, his place in front of God. I've loved the mystical since I was little, and the sheer magic of a boy hearing God call his name was a story I wanted to return to and dream of for my own life. Nana taped a spot in hundreds of pages at the one story that continues to speak to me.

The only bookmark besides the faded – what appeared to be maroon, but was really purple – ribbon was a thin, textured linen-like marker from Cadmus Books. The bookmark claimed, "Made stronger to last longer," and appeared to be from E.M. Hale and Company publishers. I looked them up and found out that they were known for Children's Books. How did she come by a bookmark from a publisher in Wisconsin? Did it come with a book order? Where did she buy her books? As an avid reader, I'm continually shopping for books. I try my best to support local, independent bookstores. I wonder if she had a favorite bookstore, or did she have a collection of favorites like me? This bookmark with its unknown origins held pages 808 and 809 in waiting. Isaiah 44 – 46, but mainly chapter 45. Did this chapter of the Bible have deep meaning for Nana? Was it a chapter being studied in Sunday school? Perhaps the congregation was reading through Isaiah? Did she try to read the Bible from start to finish like I did and only got this far, farther than I got?

Joan Didion in *The Year of Magical Thinking* talking through her moments of loss and grief states:

> Survivors look back and see omens, messages they missed... They
> live by symbols. [. . .] One day when I was talking on the telephone

Ruth in the garden, June 16, 1935. Same shoes as the graduation photo.
I imagine this is after graduation in the fancy dress she made.[10]

> in his [her husband's] office I mindlessly turned the pages
> of the dictionary that he had always left open on the table
> by the desk. When I realized what I had done I was stricken:
> what word had he last looked up, what had he been thinking?
>
> By turning the pages had I lost the message? (152-53)

I, too, found myself looking for omens, symbols. Frantically, I read and re-read Isaiah 45 for a message. Verses 2-3: "I will go before thee, and make the crooked places straight; I will break in pieces the gates of brass, and cut in sunder the bars of iron: And, I will give thee the treasures of darkness, and the hidden riches of secret places, that thou mayest know that I, the Lord, which call thee by thy name, am the God of Israel." Could I find something here? The going before, maybe. The spiral of time linking me to ancestors; moving the past forward into the present. No, the chapter where the King of Persia, Cyrus, is chosen by God to show that the Lord is in control as Cyrus and his army conquered Babylon means nothing to me. I didn't find a magical link to Nana.

[10] My aunt's statement about Ruth not being able to get a job in her field after she graduated once again brought up the WHY – oh, not the why of WHY did she not get the job; that

was easy. She was a woman trying to take on a man's job in the 1930s. I know all about trying to navigate a man's world; I have sought a career in academia, I know. But the why that came up was – why in 1935 did Ruth decide to go against all the norms and go into the science field – the science field – as a woman in the mid-30s? Let that sink in a moment. In 2015 we are desperately trying to get girls interested in the sciences because society has repeatedly told them they are not good at it, yet Nana in 1935 followed her passion and went into math and biology with a vision of teaching high school – again, high school. Women rarely taught science and math in high school, as Ruth found out when she went looking for work. So why – why did she do it? Why did her parents support her in this decision? These were the questions swirling around in my head. Not that I begrudged her of those decisions, more so because I admired her for them, and wanted to know more about her – this woman who took on a male-dominated field in a time when women were barely graduating from high school. In a recent email, my aunt informed me: "I was aware that she could not get a teaching job she wanted when she graduated from college because she was female, and that she graduated from college in 1939 when many of my friends' parents had not even graduated from high school. She was the only working mother in our neighborhood." Ruth transformed before my eyes into a feminist icon. The only working mother in the neighborhood.

Holy shit, I had a bad ass Nana – and I didn't even know it.

I was raised so differently than she was, and she was born 56 years before me. I didn't come to my feminist senses until I was in college – as a non-traditional student in my late-20s. I was already married by this time. I remember sitting in one of my college classes, and my professor asked for a show of hands of who thought of themselves as feminists. Only one woman raised her hand. (It wasn't me.) She looked around at the rest of us and said: "Come on!" Oh, I believed in equal rights and women having a voice. Yet, there was something about the term feminist that still haunted me from my childhood, enough that I didn't know if I was allowed (yes, *allowed*) to raise my hand. It is a term many Christians are raised to abhor, because, let's face it, Christianity, as we know it, is rooted in a patriarchal system. After that class, on my drive home on the dark, windy roads through Pennsylvania farm land, I came to terms with my feminist self. That moment started the path of my separation from how I was raised. Now I wonder how my view on feminism may have been different had Ruth been alive. Would it have taken me 29 years to realize that I was more than just a vessel for someone else's – a man's – vision? Would I have seen my own value? Maybe I would have been more open to a career and building a space for myself in the world earlier.

Secretly, I imagine conversations we may have had if she lived. Talking about women, the glass ceiling, education. I might have walked around the garden with her talking about what I was reading at school.

Only once has Ruth joined me in a dream: One night in Paihia, New Zealand/Aotearoa, she stepped into my nighttime world. Geoff and I were there on our 20th wedding anniversary, returning to the place where we went on honeymoon. In my dream, Nana, Aunt Pat, and I talked about university and our "I wish…" moments. It seemed so real that it took a few moments upon waking to realize it was all just a dream. The humid air enveloping me was Pacific air and not the Pennsylvania summer where I sat with Nana under a tree. It could only ever be a dream. I don't even know what Ruth sounded like, let alone her mannerisms.

SeaTac Airport waiting for my flight to Philadelphia. Sustenance to
see me through the journey and my quest to find Ruth.[11]

---

[11] Ruth: a smart woman who attended college and was the only working mother in the
neighborhood. Her mother Mary, my great-grandmother, also worked. Mary taught school,
she tutored, and she worked as a seamstress at a women's dress shop called Anne Toni's.
When her husband died, she bought a house in Collegeville where she rented out rooms to
students. From what I've been told, she sounds like a sassy lady who wasn't afraid to share
her mind.

I had a trip planned to go back to Pennsylvania. I wanted to spend time with my parents,
because they were unable to make the trip out West owing to health issues. Any time I want
to see my parents requires me to journey to Pennsylvania. I accept this since it was my decision to move away. I think we all wish they could come visit me in Washington from time
to time, but when air travel isn't an option for them, it becomes my duty to be the prodigal
daughter. Therefore, trips home often start with a goal of seeing my parents. However, I
now had a new mission for this trip. I wanted details – a closer look at the spiral.

Reading on the plane. Seems eerily fitting as I head home.[12]

---

[12] "My husband likes my family but is uneasy in their house, because once there I fall into their ways, which are difficult, oblique, deliberately inarticulate, not my husband's ways." – Joan Didion, "On Going Home"

I read that line and clarity washed over me. Earlier in the year, I wrote an essay about my unease of returning to my husband's family home in New Zealand – my sense of insider and outsider, and how that will always be my position within the Beardsall family. The same is true from my husband with the Helm family. Our ways are different, and I revert to some of my old ways when I am home. Considering that, it was nice to be on my own this trip; however, at the end of the day I was missing Geoffrey, my sounding board and my refuge from feeling displaced at the same time as feeling at home.

Home: it is a complex world/word for me. My sense of place is rooted in Pennsylvania, but it is not the place where I feel understood. Instead, I find understanding within myself when I'm in Pennsylvania as I watch and listen to those around me; I am able to connect the dots to some of my own behaviors and habits, linking myself back to moments of time and mannerisms linked deeply to place. However, I have found a space in Washington where I fit in and I'm amongst other people who think like I do. But oddly enough, my real sense of place sits in the South Pacific (see note 2). My roots span the globe, so why shouldn't my sense of place be expansive as well?

This trip was all about connecting to my roots. Reaching back into the past to see where I'm going – the spiral to help define me. I needed to dig deep within the soil of family memories to find what was hidden and forgotten.

Mary W. Detwiler watched over me as I slept in my parents' guestroom.[13]

---

[13] As I unpacked that night, I realized that this may have been the drawer Ruth used. The bedroom suite that is in my parents' guestroom was once in Nana and Pop-pop's room. The top of the armoire houses pictures of my grandparents and my great-grandparents, my history and heritage looking back at me. I noticed a picture of my great-grandmother Detwiler (Mary). I wondered if that is what I will look like when I'm old. Was the clear glass really a mirror? I can see my eyes in her, but there are differences, like our ears.

Elizabeth DeLoughrey, in her work following the thread of how the spiral moves in literature and narrative, states, "Rather than posing a facile advocation of time that repeats itself unchangingly, something that we might call 'cyclical time,' . . . the spiral, like *kōwhaiwhai* (Maori painted scroll ornamentation), *signifies repetition with a difference*" (68).

Surrounded by Nana's furniture I drifted to sleep listening to the katydids, a summer sound that lulled me to sleep for decades, and settled in to the fact that I had returned.

Grandmother Detwiler and Great-Grandmother Weikel in the garden.[14]

---

[14] Colleen and I worked together on three books and continue to help and encourage each other in our writing lives. We met in graduate school at Lehigh University. Our Pennsylvania-German heritage, work ethic, and our love for literature united us. Plus, we were a couple of the few students in the English graduate program who were married, juggling both school life, work life, and married life.

Spending time with Colleen fills my soul and helps me remember who I am. It was not surprising that for our hour walk in the park we discussed our current projects. I shared with her my quest to know and understand Nana and my great-grandmother – how these women, smart women, were advocates for education and learning. I felt as if I was linked to them in a deeper way than genetics. With each step and tidbit of information, Colleen started to share in my excitement and curiosity. I told her how the night before I had looked through all the old family photo albums from both sides of my family. My parents had gathered the albums from their siblings before I arrived, and they were waiting for me to delve into the past. With each page-turn I moved further and further around the spiral of my family, faces echoing through the generations. The crossed arm stances of Pop-pop are mirrored decades later in a photo of my brother at Aunt Pat's house.

My first inclination was to investigate the sewing and sewn aspect of my heritage. I found out Dad's father was a tailor and fitter for most of his life, as was his grandfather, who most likely apprenticed for this career in Germany. Nana was a great seamstress. And Great-Grandmother Mary also made her living as a seamstress. I was trying to link together my history with this idea of fabric and the combining of pieces to make a complete garment (see note 3). I was flinging around metaphors and trying to weave it all together. However, it wasn't until I started talking about Mary that I could feel the pull deep inside me.

E. Leroy and Mary Detwiler.
He was quite a handsome man; I can see why Mary married him.
My mom thought the "E" stood for Edward and she named my brother Dwayne Edward.
Turns out the "E" was for Edgar.

E. Leroy and Mary's home where Ruth and Marion grew up.
A farm house situated on the edge of Ursinus College campus.

Ruth Anna Detwiler. The only photo I have of Nana at a young age.
Mary Janes with knee socks is still one of my favorite fashion choices.

Young Ruth showing off her new haircut. So trendy.

Skippack Public School, 1913. Not what we thought. [15]

I started telling Colleen about how my great-grandmother had a house in Collegeville near Ursinus College and that she rented rooms out to college students. Mom told me the story that Mary would only rent to boys, because girls had a habit of sneaking boys into the house, and she didn't want to deal with that nonsense. There was something about this woman that intrigued me. She seemed witty and sure of herself. She also raised two girls who attended college and became career women. Mary took the bus from Collegeville to Norristown to work each day. Who was this woman? I needed to know.

Colleen paused and turned to me and said, "I used to teach in Perkiomen Valley School District, I know Collegeville. Where did your great-grandmother live? Is the house still there?"

I had no idea. I had never been there. I pulled out my phone and texted my aunt. Her reply: *6th Ave; 30...32?*

"I am going to take you there," Colleen said matter-of-factly. This is what I love about my friend – her willingness to join in on the quest.

[15] Coffee in hand and an egg and cheese bagel resting on my lap, Colleen and I left Sellers-ville to seek out the past of women I've never met. Driving the windy Pennsylvania roads flanked by fields of soybeans and corn, I traveled into the past in search of clarity.

Ruth, after unable to get a teaching job in the high school, worked as a statistician in Phila-delphia. She also worked as a substitute teacher for awhile. Later she taught math for a year at a junior high school, but she hated it. Finally, she started teaching fourth grade (see note 4) in Trappe, and this is where she stayed until she became ill and could no longer teach. Colleen promised to take me to the school.

Could this be the school where Nana taught Fourth Grade?

Trappe Elementary School, 1957 [16]

We arrived at Skippack School and took some photos. I tried to imagine Nana here in this place. It was difficult. The school was sold to the local 4-H and then sold again to a preschool. I sent Mom a photo to confirm this is where Nana taught fourth grade. I had just given my mom my old iPhone that morning, and I explained to her how it worked. I waited in hopes that she would be able to figure out the phone without me there. Finally, her reply came back: *No*. We were at the wrong place.

[16] We arrived at Trappe Elementary, but it looked too modern. I was going to call my mom again to confirm, but then I spied a large cornerstone on the building announcing 1957. That would be about right and fits into the timeline of when Nana would have been teaching. Taking photos, I called my mom again to confirm the location. She said we were at the right place and that across the street was the Wismer farm – or what was the Wismer farm in my mother's memory of land markers. I listened as Mom told me about my great-grandmother's sister, Annie. She married Charles Wismer, and they owned a farm in Trappe on Route 113.

Hau'ofa discussing the way that place, land, is part of the cycle , and I'd argue the spiral, of understanding ourselves and our past notes:

> It is right there on our landscapes in front of our very eyes. How
> often, while travelling through unfamiliar surroundings, have we
> had the experience of someone in the company telling us of the
> associations of particular spots or other features of the landscap

30 6th Avenue, Collegeville, PA[17]

traversed with past events. We turn our heads this way and that,
right ahead in front of our eyes we see and hear the past being
reproduced through running commentaries. (73)

Here. This place. Fourth graders once sat in a classroom with Nana. Shifting between my
fourth-grade experience and their fourth-grade experience, I wondered what happened
between these walls. What was Nana like as a teacher? A clairvoyant once told me: "You are
wonderful with children and often very patient. You expect children to be well-mannered
and respectful just as you expect adults to be the same. You have a way of letting children
be themselves with boundaries."

Accurate, I thought. It is surprising to me that I once wanted to be a mother, and I used to
teach preschoolers. My expectations of children rarely match reality. I don't think I could
handle the growing confidence and silliness of ten-year-olds all day long. Children on the
cusp of adolescence, fourth grade is integral to their development. How did Nana do it?

[17] We drifted by the old stone buildings and tree-lined streets of Collegeville until we
reached the entrance to Ursinus College. We stopped by the stone archway, the wrought-
iron arch with *Ursinus College* in gold lettering announced the entrance of the college where
Ruth earned her degree. Colleen made a right onto a beautifully tree-lined 6th Avenue.

"Wait," I raised my hands in question. "Sixth Avenue leads to the archway and the entrance
of the college?" I had no idea it was this close to the campus, nor that it was the most pic-
turesque street in Pennsylvania I had ever seen.

Soon after turning down the street, we arrived at our destination – 30 6th Avenue: the
home of Mary W. Detwiler, my great-grandmother, the woman who encouraged her two
daughters to get college degrees. What I found was a large white house, a duplex, a twin –
like the house I lived in now (and like our first house in Pennsylvania). It intrigued me that
it was a duplex. I have never lived in a single house. I grew up in a row house, a standard

The Tree[18]

type of house in the East Coast of the United States of America. In fact, our row house was the only one on our street. It was a section of four homes that used to be trolley houses (see note 5). It was moved to a quiet street in Quakertown. Most of my young adult life I lived in apartments. When my husband and I bought our first house, it was a duplex near Rich-landtown, Pennsylvania. And now in Bellingham, we also own a duplex, or twin as they are called in PA. My home choices mirrored my great-grandmother's – should I find meaning there? Maybe not, but I can't help but think about it. Duplex. Twin. Mary. Me. I'm too good at overthinking, swirling moments around and around in my head.

[18] As Colleen and I walked down the street taking photos, the street was quiet and shaded, I noticed a dogwood tree in the middle of the front yard – a dogwood, just like the one we had in our front yard growing up. When I sent a photo of the house to my mom to confirm the location, she mentioned that it was because of this dogwood tree that she wanted one in her front yard.

Looking at the glassed-in front porch, I wondered what it would be like to sit in that enclosed space and look out on to the street watching college students walking to and from campus. I thrive in the higher education world. I love to see students on campus and feel the energy in the air that they bring with their desire to learn and to discover, and their often new-found freedom. I have never lived that close to a campus, but if I wasn't working at Western Washington University, the next best thing would be to live close to campus, so I could still drink in the college experience. I wonder if this is why Mary bought this house so close to the campus. Her married home and farmland with E. Leroy linked her to the college property, and now her widowed home was walking distance from the entrance. Was she in love with education and learning like I am? Did she fill her soul by watching others

Walkway to Mary W. Detwiler's home in Collegeville.
And of course, there is a dogwood tree in the middle of the front yard.[19]

learn? I can only assume this is the case.

[19] I wonder what she would say to me if she were here now. Would she be happy that I've started seeking information on her life? Would she welcome me with a cup of tea and a slice of cake? I don't even know how she welcomed guests.

Mary died two weeks after I was born. She sent me a silver spoon when I was born. My mother sent her a thank you card with a picture of me. The card arrived the Saturday before she died. Uncle Doug was taking class at Ursinus College at the time, and Aunt Barbara would go to Collegeville with him to visit her grandmother. She was at Mary's on Monday. Mary was excited to receive the picture of me, and she showed it to Aunt Barbara. Mary died the next day.

Mary was waiting to go out to lunch with a friend, which, according to my mom, she loved doing. I can only draw more of a connection with her here, because that is my all-time favorite thing to do with my friends. When the doorbell rang that Tuesday morning, she

I am here. I have arrived at your home, Mary.[20]

---

thought it was her friend, but it was someone from the church bringing the potato doughnuts she ordered for Fasnacht. That means it was the Tuesday before Lent. She walked from
the front door to the dining room for her money and never came back. The lady delivering
the doughnuts grew concerned and called out for Mary. When Mary didn't answer, the
lady went into the house and found Mary slumped over a dining room chair. She died of a
massive heart attack.

At least, I believe, she was happy when she died. She was getting doughnuts and on the
verge of spending an afternoon with her friend. And just the day before she shared her joy
at seeing a photo of me.

[20] I stopped in front of the house, posing next to the front walk and the lamp. I'm thrilled
that I have a friend willing to not only join me in the adventure but insisted on it. It is rare to
find a friend willing to go to these lengths and who would care about a project I'm working
on. Colleen has started calling me Gyre Girl – my obsession with the spiral of time seeps
into my writing and my conversations. I wonder if Mary's friend was like this. Would the
woman she was waiting to have lunch with join her on a quick tour around Montgomery
County to find a house of a long-dead relative? I hope so. I hope Mary had a friend like
Colleen, someone who challenged her intellectually and someone who understood her so
well to know when she needed an extra push. Colleen provides the fearlessness that I need in
my world. The photo was taken and the loss started to settle down in my stomach. I found
her house. However, I still didn't know anymore about her. While this house and its location
made me think we are more alike than I realized, I was still lost. Left with questions and

At the end of 6th Avenue, only a short distance from Mary's house.[21]

Augustus Lutheran Church, Collegeville, PA

longing to know more about her, but having zero access to the information. I was here, but not able to really access the here that I needed. The here I sought was in the past, but the here where I stood was in the present (see note 6).

[21] I wanted to take a photo of the Ursinus sign – the archway and entrance to the college that Ruth must have walked under countless times. As I stood in the middle of the street after I snapped the shot, Colleen looked over to our right. I followed her gaze. A cemetery. We both thought the same thing. I texted my mom to find out where Mary was buried. It would be too easy if it was just right next door. My mom said that Mary was not buried there, but at Augustus Lutheran Church on Main Street.

My mom didn't remember where the grave was, but said it was in the middle. *The middle of what*, we thought, *the middle of the whole cemetery, the middle from the church?* We parked at the church and walked to the middle and started heading into the cemetery.

E. Leroy Detwiler's glasses and pocketwatch; Mary W. Detwiler's compact.[22]

I called my mom to find out where it was, and she said, "Oh, I'm not sure. You were two-weeks old when I was there last."

[22] That night I sat in my parents' living room with a bowl of pretzels, one of the only foods a really crave from Pennsylvania, and flipped through the pages of the photo albums. I was searching for photos from inside Mary's house. I needed to see her and see the place she lived and its space. I value my space, and everything I place in my home says something about me, whether it is the old typewriter I have on my – Mary's – library table or the retro 1970s dinner plates in my kitchen. My dining room table always has flowers on it. Did my great-grandmother have flowers in her house? Did she see gardens as a source for food or did they also provide delight and beauty, too?

I didn't find any photos from inside Mary's house, but my mom shared with me that she loved the house and dreamed of living there. My mom said that she and Dad talked about buying it after her grandma died, but my dad's work was in Quakertown, and he was unable to find work in Collegeville. So, they didn't buy it. I heard the regret in her voice as she talked about the house.

However, my mom's eyes lit up when she talked about her grandma. Mom said, "Grandma always set a really beautiful table. Flowers, good china, and each person had a salt cellar at their place setting. Sometimes we would have little sharp knives at the table. They were used for cutting apples." I think I would have enjoyed a Sunday dinner at Mary's. My mom continued, "One of the things Mary always made when we had dinner were cinnamon apples. She would peel apples and make a syrup of melted cinnamon hearts, sugar, and water. She soaked the apples in the syrup, and then served them on a bed of lettuce with a big spoonful

Gottshall family picnic with Ruth keeping a watchful eye on
her girls Barbara and Marilyn sitting at the table.[23]

---

of homemade whipped cream in the middle."

[23] My aunt continued to tell me that she was a teenager when her mother died, and that she
was more concerned about herself at the time and not the person her mother was, which I
can fully understand. It is the same reason I am asking these questions now in my 40s and
not when I still had access to more answers. I could have easily talked to Pop-Pop about
his wife. I could have been given the answers about her roots. When I asked my mom if
Mary had a Pennsylvania-German accent, she said no. So where did they come from if they
weren't German?

With the photos of the women in my life securely in my carryon, I boarded the plane in
Philadelphia with more questions. I didn't get to know Ruth or Mary on this trip. Instead, I
was given glimpses into their lives, but no answers. There was a big question that still haunted me, the one that really started me on this journey: Why college? Why did this woman
encourage her girls to go to college? And why did that not get passed down to my mother?

College was a foreign concept in my house. My brother knew he was going to be farmer
from the time he was a toddler. My sister wanted to be a wife and mother. She took some
business classes, but never really indicated any career goals for herself. I felt that I would
someday go to college, but I didn't know what I wanted to do with my life. I traveled the
world, moved to New Zealand, and late in my 20s, I decided I would follow my love of
literature and stories.

I continued to transition from the past to the present and back toward the past. The past
had a vision of my future.

Mary W. Weikel (Detwiler) and Annie Weikel (Wismer)[24]

Mary had some crazy superstitions: my mom has always talked about Mary and the number 13. Mary would eat in the kitchen when there were 13 people at the table. It seemed a bit excessive, but it was her belief, and she was dedicated enough to eat dinner alone instead of subjecting herself to whatever it is that would happen sitting in a room of 13 people. Maybe she was on to something about the number 13. Did her obsession and fear speak into the future? My future? My brother would die on March 13th in a farming accident. Did Mary have a premonition about the number 13? – the future reaching into the past. Did the number chill her soul like it does mine now (see note 7)?

I wonder what she thought of her daughter, Ruth, being born on the 13th. Did that linger somewhere in her mind when she lost her oldest daughter to cancer?

My mother would lose her firstborn on the 13th.

So maybe that fear was founded. A number that would haunt the family for generations.

[24] Reading Roland Barthes *Camera Lucida*, he discusses the quest for finding his mother in photos, his desire to recognize her. It made me think about how I have looked for myself in the photos of Ruth. Barthes states, "According to these photographs, sometimes I recognized

a region of her face, a certain relation of nose and forehead, the movement of her arms, her hands. I never recognized her except in fragments" (65). I felt like this is how I was seeing myself in Ruth. I was a fragment of her. Contemplating this need to recognize ourselves in family photos, I remembered I used to have a photo of two young girls. I thought it was of my paternal grandmother, but now with this new connection with Ruth, I wondered if my assumption was correct.

The photograph was still packed away in a plastic container in my closet. It had been in the box since we moved to Washington. Pulling out the suitcases and other plastic bins, I got to the one that contained framed photographs. The photograph in question was in a white ceramic frame with delicate pink flowers painted around the edge. I bought the frame when I was in eighth or ninth grade. Even at a young age, I was an obsessed Anglophile who leaned towards the Victorian Era. I thought the frame with the old photo would go perfectly with my Victorian display in my bedroom. The frame and photo sat on a doily, made by Nana, with a teacup and saucer, a pewter dish with pink and peach potpourri, and a pair of white gloves. A teenager's collection of family heirlooms and rummage sale finds. I looked at the photo, which I had not viewed in over four years. It was of two girls with their foreheads tilted in and touching. One of the little girls wore glasses and looked a bit older. The other girl had wavy hair, like a stylized ocean cascading over her shoulder.

Was this the picture about which Uncle Bill said I looked like the girl with wavy hair? I think so. I opened the back of the frame, still believing it was my paternal grandmother. I was shocked when I read *Mary and Annie Weikel*. This is Nana's mother. This is my great-grand-mother. I was just at her house in Collegeville two weeks earlier.

My world started swimming. I started pulling at memories of when I received this picture. I was pretty sure it was the result of Uncle Bill saying I looked like her, and I was given the picture. Taking the picture back to the living room with me, I continued reading Barthes. I reached this passage: "But more insidious, more penetrating that likeness: the Photograph sometimes makes appear what we never see in a real face (or in a face reflected in a mirror): a genetic feature, the fragment of oneself or of a relative which comes from some ancestor . . . But this truth is not that of the individual, who remains irreducible; it is the truth of lineage" (103). I paused as I remembered that it was not the photograph of the young girls that used to be on display in my teenage bedroom that Uncle Bill linked to me. There was another photo.

Mary W. Weikel (Detwiler). I finally found her.[25]

---

[25] The gyre of memory surged, bringing the past to the future and looping back again.

I started to see fragments of it in my mind. It was a small sepia image on a cardboard display mount. It was of a young woman, with a lace collar and her hair up in a delicate bun. I had it in my basket of photographs. This basket had traveled around the world with me. I knew the photo rested in a manila envelope in the basket, an envelope I put it in when I moved to New Zealand.

My heart raced as I turned it over... in pencil was written Mary Weikel. My great-grandmother. This was the photograph Uncle Bill pulled out of the family photos we were looking at one Christmas at Aunt Barbara's. He said then that I looked just like Grandma Detwiler. For some reason I always thought it was Gottshall. At the time, I didn't see myself in the photograph. I did see fragments of myself, but not clearly. I remember the night I took the photograph home with me, I stood in my teenage bedroom, teal and pink wall paper, childhood stuffed animals on bookshelves, and lifted my hair into a fist at the top of my head holding the photo next to me. I shrugged my shoulders and scrunched up my nose, a facial expression my family often associates with me, (they called it my piggy snoot – nostrils reaching forward) and let my curls fall back down my shoulder. I tilted my face from side to side. Maybe the nose, but my piggy snoot appeared again, and I shook my head. I thought they got it wrong, a sepia deception. I put the photo with my collection of faces from the past.

Now when I compare her face to my face around the same age in photos, I see how very similar we are, even though I didn't see it at the time. It is funny how photos show us things we don't see in the mirror. It is clear that it is from the Weikel line that I carry my dominate features. I finally belonged. The spiral returning. So many times it has been easy to see how my sister and my brother fit within this family. It all made sense now. I look like a line of the family, which was lost to us when Nana died in 1970 and when Great-Grandmother died in 1973. No wonder I didn't see myself in the family members that gathered around us at family reunions. No one could compare me to the people in the circle of lawn chairs at family reunions or around the candlelit table over Christmas dinner. I wasn't even alive when Ruth died, and I was only a few weeks old when Mary died. No time to see a mirrored face looking over the bowl of potato salad or steaming platter of corn on the cob.

This discovery of my resemblance to Mary only served to ignite my desire to know more about her once again. I put Barthes aside and started searching for her online. I had tried before but failed. I was able to find information about E. Leroy, her husband, but nothing on her except in relationship to him. I was searching for her as Mary W. Detwiler, but then I realized that maybe I needed to search for Mary Weikel to try to find more about her.

A search of Mary Weikel brought up a whole new list of findings. I found a Mary H. Weikel as a student of a West Chester Normal School. This piqued my interest. College, a teacher's college. Could this be where my great-grandmother got her education? If so, maybe this is why she encouraged her girls to go to college too. But the Mary Weikel I found was Mary H. Weikel, and she went to the Normal School in 1918, which would mean Mary would have been 27 – too old to be in the school, and she would no longer have been Weikel by that point. And the photograph of Mary with her sister listed her middle initial as W. This got me thinking, what if I looked up Mary W. Weikel *and* Normal School? That search brought up three search results and they were all for Ursinus College dates ranging from 1909-1910 (she would have been 18), 1910-1911, and 1911-1912. I was shocked. I found her and linked her to a college. It was unclear from the records if she was in a degree program or just taking courses. But she was a student at Ursinus College.

I decided to plug in her sister's name and normal school in a search, and sure enough, Annie Weikel attended West Chester Normal School. Not only did Ruth and her sister, Marion, go to college, but so did Mary and her sister, Annie. One going to Ursinus and one going to West Chester... as if this was the pattern for Weikel women. Reverberation of each other.

Ursinus admitted its first female student in 1880. According to the college's website: "The Pennsylvania Female College was located down the street from Ursinus. When it closed in 1880, Ursinus admitted Minerva Weinberger, the daughter of Prof. J. Shelly Weinberger. Minerva performed so well as Ursinus' first female student that the college officially opened its doors to women the next year. Minerva Weinberger graduated valedictorian of her class in 1884."

Mary went to school from 1909-1912. She married E. Leroy in 1914. Their first child, Ruth, was born in 1917. Their second child, Marion, was born in 1921. So what happened between 1912-1914?

On the high from this research, I decided to email Ursinus' registrar's office to see if they could provide any information on my great-grandmother and her attendance at the college. I emailed them on October 4, 2015 and received this reply on October 7, 2015:

Dear Rebecca,

I've found the name of Mary W. Weikel, listed as a student in the Latin-Mathematical group. In the 1911-1912 College Catalogue she was listed as a student in the Summer Session. She was not listed as a candidate for a Bachelor's Degree at that time. I later found her name in a list of "students not listed elsewhere" in an Ursinus College Register. She was listed only as a student who attended Summer Session and Saturday Courses. In the Register, she is listed as Mary W. (Weikel) Detwiler. After that, I couldn't find her name listed anywhere, either as a graduate or as a non-graduate. I checked under both names, Weikel and Detweiler. Her name is listed as Mary W. Weikel in both the 1911 and the 1912 RUBY (the college yearbook).

"Summer School was designed for persons who wished to prepare for college, for undergraduates and others who wished to pursue college courses, and for teachers who wished to fit themselves for higher grades of teaching."

This is all the [sic] I've found thus far. If I come across more at a later date, I will let you know.

Sincerely,
Carolyn
Ursinus College Archives

Mary

The X marks Ruth (front left) and Marion (far right) with their grandparents. I'm so very curious about the woman sitting in the middle like the queen of the family.[26]

---

[26] Sadly, I know nothing about Mary's parents, my great-great-grandparents. Mom and her sisters were unable to even provide me with names of their great-grandparents.

Having more data on Mary and her sister Annie (sometimes recorded as Anna) I went on a search to find out more about the Weikels. I found:

Pictured above: Mary L (born 11/1861) and Irwin Weikel (born 7/1861) – my great-great grandparents, Mary's parents. Irwin's parents were Noah W Weikel (born about 1833) and Emma (born about 1840) – Noah and Emma lived in Rockhill Township, Bucks County in 1880. This is the township where my parents live now.

Noah's parents were Charles Weikel (born 10/1803) and Catherine/Catharina Wambold (born about 1808 or 1809).

The farthest I was able to go back in my limited research was Charles Weikel's father – Christian Weikel born in Pennsylvania in 1779.

This was the first time my mom and her sisters found out about the Weikels. They had none of this information prior to my research.

2/22/43

February 22, 1943 - Mary[27]

---

[27] In her book *Native Land and Foreign Desires*, Lilikialā Kameʻeleihiwa states:

> It is interesting to note that in Hawaiian, the past is referred to as
> *Ka wā mamua*, or 'the time in front or before.' Whereas the future,
> when thought of at all, is *Ka wā mahope*, or 'the time which comes
> after or behind.' It is as if the Hawaiian stands firmly in the present,
> with his back to the future, and his eyes fixed upon the past, seeking
> historical answers for present-day dilemmas. (22)

This is the spiral. Where the past is sought out to provide context, answers, and understanding. The future is not where the answers lie. I look at Mary standing there on the porch and desperately want to reach out. I want to be behind the camera so I can ask her about... well about anything really – her childhood, why she is standing here in a ¾ length sleeve in February, what is that vine behind her, did she plant it, what is in the bucket, where is she going? I see you, Mary. You are in front of me – literally and figuratively – what are you telling me about myself?

Spring 1942

Winter 1942

1942 for Gottshall's. New home and baby on the way. Barbara,
their first child, would be born December 14, 1942.

(R to L) Bob holding Bobby, Flossie, Marie, Christian, Earl, Ruth, and Bill.
Is Flossie wearing roller skates? It sure looks like it.

Bill and Ruth with Bill's parents Sadie and William.

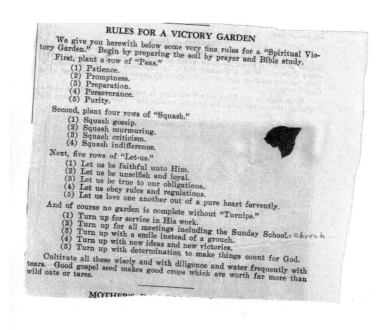

RULES FOR A VICTORY GARDEN

We give you herewith below some very fine rules for a "Spiritual Victory Garden." Begin by preparing the soil by prayer and Bible study. First, plant a row of "Peas."

(1) Patience.
(2) Promptness.
(3) Preparation.
(4) Perseverance.
(5) Purity.

Second, plant four rows of "Squash."

(1) Squash gossip.
(2) Squash murmuring.
(3) Squash criticism.
(4) Squash indifference.

Next, five rows of "Let-us."

(1) Let us be faithful unto Him.
(2) Let us be unselfish and loyal.
(3) Let us be true to our obligations.
(4) Let us obey rules and regulations.
(5) Let us love one another out of a pure heart fervently.

And of course no garden is complete without "Turnips."

(1) Turn up for service in His work.
(2) Turn up for all meetings including the Sunday School. church
(3) Turn up with a smile instead of a grouch.
(4) Turn up with new ideas and new victories.
(5) Turn up with determination to make things count for God.

Cultivate all these wisely and with diligence and water frequently with tears. Good gospel seed makes good crops which are worth far more than wild oats or tares.

MOTHER'S

Another treasure hidden away in Ruth's Bible [28]

[28] Nana's Bible sits on my dining room table, the glass top placed there to preserve the cedar planks of our handmade table reflects the light and makes it appear that the Bible is glowing. Last time I opened it was to look at the spot of the bookmark. I also pulled at the purple ribbon at the time to see the space it marked: St. Mark 2-3. No matter how many times I read the passages, I grew no closer to Nana or understanding her. I closed those heavy pages and walked away. But I came back to it, flipped the pages, and movement caught my eye. I went back to that section and found a paper tucked in 1 Chronicles 21, and with the paper there was a tiny leaf. I wondered if it was a rose leaf since Pop-pop was known for his roses. "Rules of a Victory Garden." Oh, gardening. Gardening cycles me back to my childhood and my memories of helping Mom plant flowers, pick vegetables, and water plants. I never, conveniently, remember weeding. With delight and memories of my hands in the soil, I returned to the paper in the Bible about Victory Gardens, only to be disappointed. The piece is a clever play on gardening. Instead of focusing on real plants, it is about a "spiritual victory garden." Perhaps Nana cut it out because of her love of gardening, too. I hope that is the case, but maybe she liked the message. I see she wrote on it. She added the word "church" in the Turnips section: "Turn up for all meetings including the Sunday school. (& church)"

Yes, I know all about turning up for all meetings. My childhood felt like a continual movement to and from church. Wednesday nights, Sunday mornings, Sunday nights were the given. Then accompanying my parents, who were attending meetings and Bible studies.

Zealand farms.

Shudders at Snow

Carlos, although happy for the opportunity to visit farms in the United States, shuddered at all the snow which had fallen since his arrival in Lawrence County. With a pleasant smile, he explained that on his family's farm the temperature now would be about 85 degrees, mildly-tempered by a Gulf breeze.

Main crop on the farm near Vera Cruz is bananas. Other fruit grown includes oranges and avocados and the vanilla plant from which is extracted

The backside creating a connection through time.[29]

---

[29] Frustrated by the false hope of information on a Victory Garden, I flipped the page over. The first thing I noticed was the picture of the two men. A section on farmers. I'm curious now where Nana found this article. Was it one of those farming chronicles, papers, magazine that used to be found around our house, too? We used to have a stack in a basket in the bathroom. My eye scanned the page again. "Zealand farms" rested at the cut edge, and my heart jumped. The "New" left in the void – removed by the scissor blade – but I know it was there. A word my eyes are trained to see – New Zealand/Aotearoa.

The past reached out to the future. What are the chances that Nana would cut out an article on spiritual gardening and the back of the page, not regarded or valued in the cutting, would keep the remnants of the name of the country her granddaughter moved to and considers home?

Yes, as Didion mentioned, I was searching for symbols, but this seemed a little too coincidental to just walk away from it. A clear connection through time, I'd say. And New Zealand being the reason I even know about the spiral of time. A non-linear understanding of time, which I first discovered in narrative, without knowing the terminology, in novels by Maori authors like Patricia Grace.

When I was reading for the comprehensive exam for my MFA, part of my theory list included Elizabeth Deloughrey in "The Spiral Temporality of Patricia Grace's *Potiki*," where she states:

> Rather than segregating the "past time" of the ancestors from
> the "present time" of the contemporary community, Grace
> employs a spiral temporality where past and future times is
> narratively re-experienced in what she terms the "now-time,
> centred [sic] in the being." (60)

Now. Here in this moment. New Zealand pulled me back to the page clipped out of a magazine by Ruth. The koru. An opening, but also a closing. This is the heart of the koru – the

A trail of four leaf clovers through the pages. [30]

---

unfurled fern frond. A sign of birth, beginnings, awakening.

Ka hinga atu he tete-kura - ka hara-mai he tete-kura.
As one fern frond dies, one is born to take its place.

[30] A found half of a word –Zealand – sparked me to go again through the pages of Ruth's Bible. There had to be more. This was a sign, wasn't it, that I needed to look again. Pages 920-921, Ezekiel 16, I found a four-leaf clover. A universal sign of good luck. This made me think of Mom. She always looked for four-leaf clovers. I never found them, but she did. Turning more pages, I found more clovers. One leaf had separated over time, and that is when I realized it wasn't a rose leaf with the Victory Garden piece, but what remained of a four-leaf clover. It must have also been a clover remains left in the page with the bookmark from Cadmus Books.

I've looked through this Bible before and never found them. The main reason I looked at the Bible previously was for the Family Register found between the Old and New Testaments. Nana's Old Testament was filled with four-leaf clovers. I wondered if this was the space used to preserve her finds, because the Old Testament is rarely referred to since the new covenant and the birth of Christ. I was willing a theory into this moment when I found a four-leaf clover in Matthew. There was no theory. No line to follow. They were just clovers pressed into the pages of the Bible.

I posted a photo on Facebook with a caption, "Found in Nana's Bible - how long has it been there?" and my mom responded immediately. Her sisters followed:

Marilyn - *I don't know but she died in 1970 and I also know she would find them when she was hanging up the wash. I remember when I was young crawling in the grass looking for them when she showed us one she found.*
Pat - *Mom would look down at the grass and pick a four-leaf clover and I never had any luck finding them!*
Barbara - *I, too, remember mom finding 4 leaf clovers especially when we hung wash some summers. I have found a few, so far none this summer.*

I took two things away from this:

What do you think of Spivak, Myrtle? [31]

1. Nana had a keen eye. How on earth did she find four-leaf clovers while hanging up wash? Perhaps it is what made hanging up wash more enjoyable… something else to do, search for, while doing the mundane, endless work of laundry.

2. Apparently, she created a line of women who continue to seek out four-leaf clovers: Marilyn and Barbara who to find them, and Pat and Rebecca who seem to have given up to fate that they will not find one, but secretly scan the ground when near a patch of clover.

[31] It seemed odd that Ruth would just press clovers in the pages of her Bible. My mom taught us to press flowers in wax paper in a book. We would have never thought of pressing the flowers without a protective covering from the pages. Also, Mom had us press flowers in the dictionary. I suppose because it was the biggest book we had. Surprisingly, I think Ruth's Bible is actually thicker than the dictionary we had growing up.

When my beloved cat died on January 1, 2020 a friend from work gave me a few sprigs of myrtle since our cat's name was Myrtle. She suggested I press some to remember our girl. I knew exactly where I would place the myrtle leaves. Dutifully, following the former instructions of my mother, I placed the leaves in wax paper and pulled out my *Norton Anthology of Theory and Criticism*.

I am a cat person. I was the kid that begged her parents every Christmas for a cat. My mom and my sister are both allergic to cats. I did eventually succeed in getting a cat, Magee, who was my shadow until he died when I was a senior in high school. After I finished my undergrad as a non-traditional student, I went into the English graduate program at Lehigh. I had been searching for a British Shorthair cat for a couple of years after watching the film *Girl, Interrupted*; I loved the look of British Shorthairs, but I didn't know the breed until, thankfully, the movie showcased one and I could look online for information on the breed. And with graduate school in full swing, I really wanted a cat around. Something/Someone to be with me through all those hours of studying and writing. Then I found Myrtle. She was with us for twelve years and saw me through two graduate degrees. She was the best study partner. On this particular day, I was reading Elizabeth DeLoughrey's article "The

Marie and Ruth.
The awkward moments of family gatherings.[32]

Spiral Temporality of Patricia Grace's *Potiki*," in preparation for my MFA comp exams. In the article, DeLoughrey references Gayatri Spivak. I knew I had one of Spivak's books in my library from my earlier graduate degree where I focused on postcolonial literature. I decided to follow my training and go to the source. Yet, I searched and searched and couldn't find the Spivak book. I realized I must have sold it when we were planning our move to Washington State. We originally planned on shipping our stuff, and the cost was based on weight. We looked at my books and sighed. Sadly, I had already given away a bunch of books by the time we found an alternate shipping method. I digress. The only book that I have with anything written by Spivak was my Norton Anthology. I pulled the book out, Myrtle joined me on the floor. I looked over and said to her, "What do you think of Spivak, Myrtle?" She looked up at me as if to answer. Thus, the leaves to remember her rest in wax paper between the words of Spivak.

I wonder why Mom told us to press flowers between wax paper. Why do we press flowers and leaves at all? To preserve. To remember. I wonder if 50 years after I am dead someone will find those myrtle leaves in a Norton Anthology of Lit Theory and try to find meaning in the words on the page, when really those words meant nothing to me. It was the moment with Myrtle that I wanted to remember. So maybe it was not about the passages where the clovers exist that matter, it was what Ruth was doing and wanting to remember that pulled her to those pages.

[32] I wonder if Ruth felt uncomfortable in social settings like I do. Was she an introvert needing time on her own to recharge? Did she enjoy the family picnics? The holiday gatherings?

Bobby, Sadie, Mrs. Drivalbis, Marie, Ruth, Florence (Flossie). The Gottshall women with full gazes at the camera while Ruth looks toward a different horizon.

I tend to hover around the perimeters waiting for the socially appropriate time to leave without offending the host unless I'm with people I'm extremely comfortable with. I have never been at ease with small talk. I wonder if Ruth was the same. I like to think she was. In the photos of her at the family picnics, she appears slightly uncomfortable, arms crossed, as if she is trying to find her space (and hold onto herself) within this family who are now part of her story. I wonder if she sometimes felt like an outsider yet insider like I feel with the Beardsalls. Was she ever able to find something in common for them to talk about? Did she struggle to find topics that were safe like I do when I'm part of the larger family gathering?

I do have fond memories of Gottshall family reunions at my Pop-pop's. Old wooden chairs from the grange hall, playing wiffle ball with my cousins, and eating ice cream with wooden spoons. But I was a child when family gatherings just meant time to play with my cousins, not sitting around making polite talk.

My all-time favorite family gathering is Thanksgiving. I love it because we have a tradition; each year the Helms and the Cluleys get together to celebrate Thanksgiving. I have memories of waking up to the smell of turkey and potato filling and watching out the window for my Aunt Pat, Uncle Bill and my cousins Seth and Drew to arrive.

It is hard to think about Thanksgiving without bringing to mind the famous Norman Rockwell painting "Freedom of Want," which was the cover of *The Saturday Evening Post* on March 6, 1943. The woman in her blue floral dress and white apron presenting the perfectly roasted turkey while her husband looks down at the platter with adoration or hunger. All the happy, blissful faces are leaning in just thrilled to be there and to be with their family. It has been the faces that have drawn me to Rockwell's work. When I was 14, I would check out books from the library of Rockwell paintings. Since I was an art major, my parents assumed my obsession was the painting techniques, but really it was the stories I would create in my mind as I gazed at the paintings. Once again my past collides with Nana's. I had no idea Ruth worked for Curtis Publishing when she graduated from college. Curtis was one of the influential publishers in the early 20th century. It published the *Ladies' Home Journal* and *The*

Ruth in the driver's seat. [33]

Saturday Evening Post amongst a growing list of periodicals. Rockwell's work started to appear on the cover of *The Saturday Evening Post* in 1916, and he painted more than 300 covers for the *Post*. On Christmas 1988 my parents gave me two books on Rockwell. Yet no one seemed to bother to tell me that Ruth worked for the famous Philadelphia publishing company.

[33] Ruth driving. I never thought about it. But of course, she did. The other woman in this photo is Carrie. Carrie lived in Philadelphia and Nana stayed with her when she worked at Curtis Publishing. Where was this photo taken? Philadelphia? Is Ruth heading back to Mary or away from her?

Jane Flax talks about the complicated mother/daughter relationship. She claims that the child needs to move through the relationship with the mother (or caretaker) through separation and individuation. For mothers and daughters, there is this struggle to separate, but at the same time, they long for the identifying bond. In coming of age, "The girl, unlike the boy, cannot repress the female part of herself and totally reject the mother, because it is precisely at this stage that she is coming to an awareness of her own femaleness, that is her gender" (178).

I spent a lot of time reading Flax in graduate school. Mainly when I was working on a seminar paper about Little Red Riding Hood of all things. I often return to the quote above. The way mothers and daughters are linked, but also the need to separate and create the individual. Yet, through this journey of discovering Ruth (and Mary). I'm finding that the cycles continue. We try to break them, but the link to our ancestors, and particularly in the matriarchal line, are strong.

Mary – Ruth – Marilyn – Rebecca.

Repetition. Connection. Yet complete individuals with echoes of each other.

"Girls, because they are of the same gender as the mother, tend not to develop firm ego boundaries; they never completely separate from the mother. The mother often treats the daughter as an extension of herself and discourages her from establishing a separate identity" (Flax 162).

8/16/49

8/17/49

The Shore. What are you looking at? Marilyn and Ruth have their eye on a different prize while Bill sneaks a peek, and Barbara looks at the camera. [34]

---

[34] I could spend every day at the beach. I love the mystical, powerful ocean as it arrives and departs the land. When I was a kid, we used to go to the shore at least once a year. Those days were hot, sunny, with us running in the waves and playing in the sand and trips to the boardwalk for saltwater taffy. As I got older, trips to the beach would mean swimming out beyond the breakers, beach football games, and primping for the night on the boardwalk. However, the shore was always a two-to-three-hour trip from our house in Pennsylvania.

It wasn't until I lived on an island did I realize my need to be near water and how spending time by the sea replenished me. When I lived in New Zealand/Aotearoa, I made it a ritual to walk down Queen Street to the waterfront every day. Even today it becomes my deepest joy to stand on the shores of New Zealand/Aotearoa and drink in the place where mystery dances in the water. There is something magical about the South Pacific. It is hard to put my finger on it, but it makes me appreciate the need for gods and goddesses (see note 8). When Geoff and I talked about moving away from Pennsylvania, we knew we had to live by the water. It was so hard for both of us living in landlocked Pennsylvania.

I know Pop-pop used to love going to the beach. I wonder if Ruth loved the beach. Did she like it for the sun and sand? Or did she love it for the sea, like me? I know all three of her daughters love going to the beach and find joy walking the beach letting the ocean tickle their toes. I'd like to believe that my love for the beach and the water is rooted in my connection to Ruth, a place to physically witness and contemplate the movement of time as the ocean and land combine.

Nana and Mom.

Christmas Eve, 1962. Ruth's piano making its presence known.

Bill and Ruth's silver anniversary. January 1, 1966. In
four years and six days, Ruth will be gone.

The family that reads together... is there a saying that starts like that? If not, there should be. We lovers of stories inherit the need to find sustenance in words.[35]

---

[35] When Mom's older sister, Barbara, heard I was seeking out photos of the inside of Mary's house, she pulled a small stack of photos from a box in her attic and sent them to my mom. This collection is from Christmas 1970. Mom's note with the photos:

> Look at those built-in bookcases. I loved them. By all the bookcases you can see Grandmom loved to read, too. We are all sitting in the living room, but you can see the dining room and kitchen.

I have been obsessed with books from the time I was little. We had two sets of shelves in our living room framing the large window facing the front porch: two shelves dedicated, for years, to my childhood favorites. I have clear memories of pulling books off the shelf and toddling over to Mom, asking her to read to me. I see Mary also liked to have her books behind glass doors. I recently remodeled my library and added in shelving with doors. Another enthrallment – glass doors – that and a ladder have been on my library dream list since I was a teenager.

Mary opening her Christmas gift. It looks like a music box. (That's my sister in the red plaid skirt and Aunt Pat in the red plaid dress and black boots – oh, I do love the fashion of the early 1970s.) [36]

---

[36] I noted the date on the envelope containing the photos from Aunt Barbara: December 1970. Nana died on January 7, 1970. These photos show the first Christmas without Ruth. Having lost a sibling when I was 19, I know what that first Christmas without the expected, and previously constant, presence is difficult. Aunt Pat is 19 in these photos. We both experienced loss at 19. I lost my brother, Dwayne, to a farming accident. Aunt Pat lost her mom, Ruth, to cancer. I didn't realize we were the same age moving through grief of a close family member. The closest, really.

Mom (in yellow) with her sisters and cousins
circling around the matriarch, Mary. [37]

---

[37] Mom wrote: *The pictures marked June 1969 were from Grandmom's birthday. She took us all out for dinner. You can see out the window into the sun porch. (I loved that porch. It didn't have heat in it, but I remember putting on my coat and sitting out there in the winter.)*

Three strong, courageous, college-educated
women – Ruth, Mary, Marion

The porch with its huge windows was one of the first things I mentioned when looking at the house with Colleen. It appeared to me like a place I would love. So, it makes sense that it was a favorite spot for my mom, too. I like the outdoors, but not the outdoors, if you know what I mean. I like the fragrant air filled with a mixed bouquet of flowers, trees, and herbs swirling together to make the scent of the outdoors. I love the soft breeze, and the sun gently kissing my skin. But I do not like bugs, especially Pennsylvania bugs. The idea of an enclosed porch was something I dreamed of since I was a little girl. And here... here Mary had the most perfect, window-lined porch overlooking a tree-lined street in a college town. I now understand. I recognize my tendencies and my likes. I am an echo of Mary.

Aunt Pat showed me this picture of Ruth on what would have been Ruth's 101st birthday. We believe it was taken the first year she taught fourth grade. It is my new favorite.

Oh dear lord, the pattern. I love it. I can see myself in that smile.
I know what she is thinking because I've thought the same thing.[38]

---

[38] Still on my quest to learn more about Mary, I sent an email to Mom and Aunt Pat to ask if their mother shared any mannerisms with their grandmother. I was trying to continue to connect the dots from Mary to Ruth to me. I wanted the link to be stronger and to confirm that I belonged to them. I never knew them, but I already admired them and their bold way of approaching the world: having careers and being part of women in higher education. The response I got from Aunt Pat was not what I was expecting:The response I got from my aunt was not what I was expecting:

> Mom died 44 years ago, Mary died 41 years ago. My memories of both
> are vague at times. I have clearer childhood memories than I have
> memories from my teen years.
>
> I have always held my mom up on a pedestal due to her accomplish-
> ments. As a skilled seamstress: Lydia [my Aunt's grand-daughter]
> has the tulip quilt on her bed. So this week I talked to her about
> this quilt that my mom made 80 some years ago! I have wonderful
> memories of standing at the piano singing with my sisters as my
> mom played "I saw mommy kissing Santa Claus."...She was born 98
> years ago on 10/13/1917.
>
> The pedestal was very high and still is!

This realization that my Aunt remembered Ruth mainly from her childhood years and not her teenage years made me think about my Aunt's comment on Facebook after I posted a picture of myself in Atlanta. She stated: "Rebecca, I looked at your picture and saw Ruth looking at me!" Her emailed linked to that comment made me get out the calculator. When Nana was my age Aunt Pat was eight years old and my mom was fifteen. When Ruth died Aunt Barbara was 28, mom was 26, and Aunt Pat was 19.

So I am at the age when Mom and Aunt Pat hold the fondest memories of their mother. It is no wonder that they see her in me now. I have become the catalyst to recall memories about their mother. I'm not sure how to deal with this revelation. Yet, I'm kind of proud that I'm associated with her. I don't have children and so it seems strange to be linked to someone else's childhood/mother memories.

This link to their mother from their childhood also brings on fear. Ruth died from cancer when she was 52. When they found the cancer it was all over her body, but there are some theories that it started in her colon. My cousin Debbie also died from colon cancer at the age of 44.

I asked my mom, again, about Nana's cancer. Her reply, "They didn't know where it started and they never really told us what kind she had until they found it. It was all ready in her liver. She had her surgery the day after Thanksgiving and she only lived six weeks. The surgery was to remove her gall bladder but when they went in they found so much cancer they just closed her up."

Cancer stole Ruth from me, and I can't help but feel angry about that. In my quest to learn more about her the loss sinks deeper. What I wouldn't give to talk to her, to hear her voice, to listen to her ideas on life. I wonder what she thinks of me, and what I have done with my life. Is she proud to call me grand-daughter? I would like to think so. I just can't help but wonder if my life would have been different had she lived. Alas, we don't get to determine our place in the spiral, but in order to grow we must look toward the past for understanding and clarity. I'm thankful that Ruth and Mary are part of the past that defines me. Now that I know them, I will continue to seek them for wisdom and guidance.

I will return here. The here always remains.

Grading papers.[39]

39

Mary. Ruth. Marilyn. Rebecca.

(Mary Tyler) Moore added that the director told her, "'You know what would be a good idea, Mary? Go out in the middle of the intersection and take that tam off and throw it in the air.'"

Afterwards

November 14, 2015

Today I received in the mail a box of letters, report cards, and a high school year book. My mom sent me a box of Ruth.

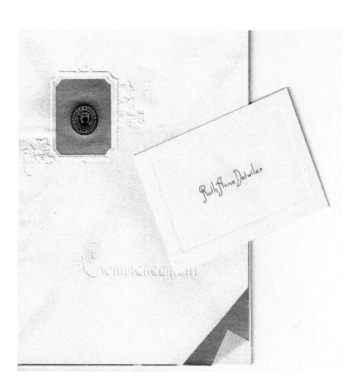

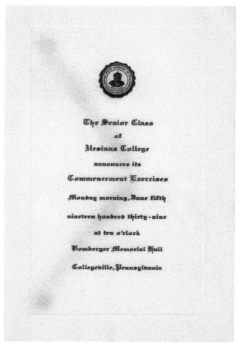

The Senior Class
of
Ursinus College
announces its
Commencement Exercises
Monday morning, June fifth
nineteen hundred thirty-nine
at ten o'clock
Bomberger Memorial Hall
Collegeville, Pennsylvania

Just before the box arrived, I was thinking about my two sides of the family. I realized that it is my dad's side of the family who are the storytellers. I recall the stories Aunt Shirley shares whenever we are together – the latest story was of her diving into the ditch and staying there thinking it was an air raid in Richlandtown. Finally, her mom and brothers had to go searching for her and found her there muddy and fearing for her life.

When I was driving home, I noticed the dark rain clouds covered up the mountains to the East, while the sea air pushed the clouds apart behind me, and the setting sun shone on the fall leaves dancing in the wind. I started to look around for a rainbow.

The rainbow. Dad's story of a rainbow started racing through my head. How he and his friend ran to the field where the rainbow ended looking for gold, but instead stood in the marvelous colors and watched it change the color of their skin. I have no idea if this was true, but it was his story all the same. And I'm sure there was truth in it. Because isn't that what makes a good storyteller, truth mixed in with magic?

Thinking of the stories – I started thinking about my family – the stories my mom shared with me were in books. She would read to me all the time and encouraged me to love books. On the other side, my dad didn't like to read, but he always had a story to share.

Because of my recent research on Nana and Great-Grandmother, I was wondering what this all meant. Ruth and Mary seemed to be more into the sciences than humanities, but I also know that They liked to read. Ruth had a lot of first edition novels, but they seemed to have a scientific mindset. Ruth and Mary seemed more practical and logical. I like to think that I'm a bit of a mixture of both practical and lover of the story. I wouldn't say that I am the best storyteller, like my aunts and uncles on the Helm side, but I do love to investigate and analyze literature and share my findings. I guess I do lean more to the Ruth and Mary side of things. Funny how looking for a rainbow can start me on the pathway toward more discovery about myself, my past, and my future.

I had just finished writing the above paragraph when the doorbell rang. It was the postal worker with a box from my mother. The content of the box opened the doorway to more discovery about Ruth. I laughed when I saw that she got a D in English at Ursinus, yet A's in math and science. Clearly, my love for the story comes from the Helms, but my analytical mind comes from the Detwilers.

In the box was Ruth's senior year book, The Colonel 1935.

WILLIAM HENDRICKS GOTTSHALL

"Bill"   *"That's what you think"*

"Bill" came to us from Norristown High during his Junior year. He often uses his big car as a bus when there are class functions. His most noticeable feature is his bright red cheeks which are often referred to as "that school-girl complexion."

Dramatic Club, 3; French Club, 4; Class Play (Electrician), 4.

Rosy cheeks run in the family. I remember Uncle Earl, Pop-pop's younger brother, commenting on my pink cheeks one evening over dinner in Pough-keepsie, New York. I love that Pop-pop's saying was, "That's what you think." He was always few with words, but when he had something to say, he would tell you. I often think of the quote by Rousseau when I think of Pop-pop: "Generally people who know little speak a great deal, and people who know a great deal speak little." Pop-pop was quiet, but you knew he had a clear understanding of all that was around him; he just chose not to speak about it.

RUTH ANNA DETWILER

"Ruth"   *"Oh, heck"*

Ruth did not take a profound interest in extra-curricular activities, but she made up for it by her good marks. Her ambition is to follow in her mother's footsteps and be a teacher, but she is also quite musically inclined.

Glee Club, 1, 2, 3, (Associate Secretary), 4; French Club, 4; Hockey, 2.

So, she was indeed following Mary's lead. Ruth obviously looked up to her mother and wanted to pursue a career as Mary did. And of course, it was noted about her good marks and lack of social engagement – that sounds familiar. Oh, heck Ruth. What do we make of this crazy world we live in where I am so similar to you, but never knew you?

Notes

1. Quilts made by Nana. Aunt Pat has the tulip quilt made by Nana, and she has given it to her granddaughter. All three of Ruth's daughters received a quilt from their mother: a vine of leaves made from their dresses, shirts, skirts, curtains. That is the one that rested on my parents' bed all though my childhood. Sadly, Nana was unable to finish Pat's quilt before her death. It needed the final quilting, and the ladies of Eden Mennonite stepped in and finished the quilt.

The first quilt Nana made when she was 7, in 1924. This quilt rests in my mother's cedar chest. Aunt Pat said that Ruth's first quilt was used hard; she remembers it being on her bed when she was a kid.

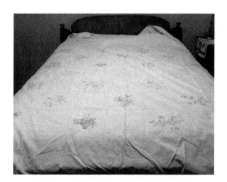

2.

New World

We arrive late. Their door an altar to sturdy, mud encased boots. On the table, shadowed by low lights, a Sunday meal - lamb w/ mint sauce, roast potatoes, pumpkin, and Kumara - A mum dressing the table for her son. The evening complete with a cup of Milo.

The room is small; we pause at the twin beds meant to house us for the night. The brightness of the dawn wakes me from a fitful sleep enveloped with the bleating of sheep. I struggle out from the duvet. I can still hear the cries of my dreams. Pulling back the faded, floral curtains I see a flock of sheep with newborn lambs in the paddock.

As I open the window, the fragrance of freesias drifts up - a fantail perched on the totara flicks his impressive tail - I stand in my new world. And I understand the fears of children.

3

Wer bin Ich (who am I)

I am from Gottshall and Detwiler;
roast beef dinners, potato filling in Franconia,
hot chocolate on cool summer mornings.
I am from gurgling water cooler,
plastic fruit, and scattered newspapers.

I am from Helm and Bleam;
pipe smoke and a tender hand
wrapped in a green cardigan.
I am from hard candy in a crystal dish;
carrots swimming in ice, cold water.

I am from tillers
of the clay soil who listen
to the summer corn grow.
I am from waiting in barns at 3 a.m.
to help a calf take its first breath.

I am from wooden pews;
Stained glass proclaims my family
name, It is well with my soul.
I am from Men-o-lan
Wilhelmina, Eden, Lou-Hagin.

I am from pacifists, hard
work and knowing your worth;
Mennonite weddings, funerals, and potlucks.
I am from Fastnachts, funny cakes, apple pies
summer kitchens and root cellars.

I am from Richland;
camp fires and lightning bugs
red beet eggs and chow-chow.
I am from citronella candles, Dutch-Blitz.
Kannst du micka funga?

I am from a family of five.
A brother who held my heart.
A sister I fought within orange walls.
I am from a father who carried me into the waves
and a mother I followed into the garden.

4.

Fourth grade is a moment in time that sticks in my memory. It has circled around my mind for years. I had a teacher, Mrs. Crouse, who was a mixture of sweet and a little bit of eccentric old lady. At least she seemed elderly to my fourth-grader, ten-year-old mind. I remember that she wanted all of her students to go to her house for Trick or Treat on Halloween. She was the only teacher whose house I ever visited. I wonder if Nana's fourth-grade students remember her the same way Mrs. Crouse lives in my memory. Does Mrs. Gottshall come to their minds when they reminisce about grade school?

In addition to my teacher, fourth grade seemed to be a turning point in my life. So much so I'd continue to write about it. Coming to terms with my own failings, I wrote this poem in 2012:

Fourth Grade

Every public Elementary
classroom has one...
The child shunned
because of his ratty clothing,
his strange unwashed odor
oozing from his skin,
misshapen teeth, uncombed hair
In my class, he was Frederick
a forceful F
imitating his buck teeth.

Mornings he arrived
smelling of syrup and sausage.
Sticky residue on his pale cheek;
dirty fingernails tried to scrape
it off. His blue and white
striped shirt sleeve
stained orange,
yesterday's pizza sauce.
Gray pant knees green
from last week's playground slide.

We learned square
dancing for the school fair.
Merciless teacher pairs boys
with girls. Beautiful children
with beautiful children. The average
haphazardly paired without
thought or care. Fred
and I received jeering
laughs and sympathetic shrugs

In his eyes I see fear.
My precarious social
position forced me to make
a face laced with disgust —
to touch the boy
my peers mocked.
Hand, no, fingertips brushed
Fred's shoulder and I realized
the singe of self-righteousness.

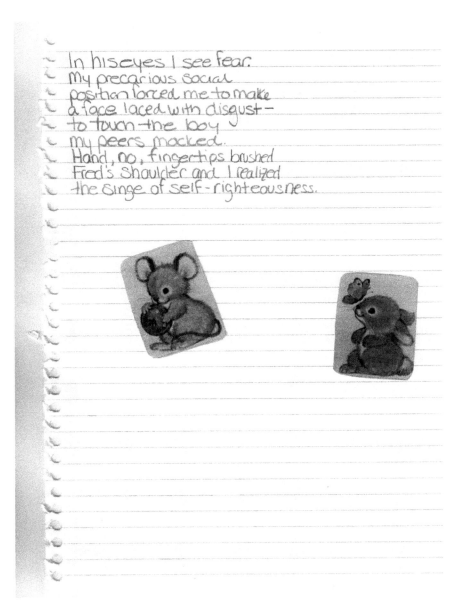

5.

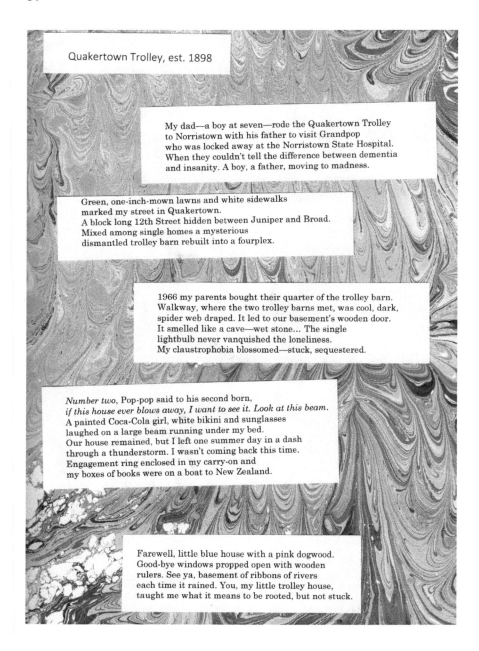

Quakertown Trolley, est. 1898

My dad—a boy at seven—rode the Quakertown Trolley
to Norristown with his father to visit Grandpop
who was locked away at the Norristown State Hospital.
When they couldn't tell the difference between dementia
and insanity. A boy, a father, moving to madness.

Green, one-inch-mown lawns and white sidewalks
marked my street in Quakertown.
A block long 12th Street hidden between Juniper and Broad.
Mixed among single homes a mysterious
dismantled trolley barn rebuilt into a fourplex.

1966 my parents bought their quarter of the trolley barn.
Walkway, where the two trolley barns met, was cool, dark,
spider web draped. It led to our basement's wooden door.
It smelled like a cave—wet stone... The single
lightbulb never vanquished the loneliness.
My claustrophobia blossomed—stuck, sequestered.

Number two, Pop-pop said to his second born,
if this house ever blows away, I want to see it. Look at this beam.
A painted Coca-Cola girl, white bikini and sunglasses
laughed on a large beam running under my bed.
Our house remained, but I left one summer day in a dash
through a thunderstorm. I wasn't coming back this time.
Engagement ring enclosed in my carry-on and
my boxes of books were on a boat to New Zealand.

Farewell, little blue house with a pink dogwood.
Good-bye windows propped open with wooden
rulers. See ya, basement of ribbons of rivers
each time it rained. You, my little trolley house,
taught me what it means to be rooted, but not stuck.

6.

Tama-nui-te-Ra

Dusk blankets
the earth in shades
of mourning. Sun

leaves an echo
in the sky. My life
pages are being turned

too quickly. A woman
holds the spine
in her hand

Using her thumb to release
the story – my narrative
flutters by her gaze.

Joshua prayed
for the sun to stand
still and the Lord obeyed.

Belief and prayer will
bring me more time?

Maui commanded
the sun to slow
down. In the flax

he snared Tama-nui-te-Ra,
with a jawbone he beat
the great glowing beast.

Wit and might will
bring me more time?

A moment to stop the spiral
to read the ritual and remnants.

- 77 -

7.

Blue

Curves twist around fields laid bare for seeds. A Blue Ford
tractor sits vacant in the field, the disker caked with clay.
Long breathe let out as memories begin. A world left behind
returns in an instant.

Shrine of diecast tractors held a boy's dream. Her tiny hands
denied access to his world. Blue eyes follow as he lifts each
tractor like a jewel to be marveled at - she watches with
impatience as instructions are given.

He names his Pipersville farm Morning Glory. . Toys replaced -
the dream realized. Dusty barns house Holsteins. She opens the
white, paint-chipped door; bittersweet smile as she breathes
in the scent

of shit, hay, and feed. The feed - for his charges - milk
producing employees. He laughs and puts on his hat. She follows
him into the cold, sterile milk room as he writes mysterious
numbers on mom's old chipped clipboard.

On a dark March night, he checks his girls before going out
with Steve. The cows recognize his voice. One last thing to
check . . .a moment only he will know. Head first fall - feed
mixer blades break his neck.

Breathless, she is lost on a road staring at a blue Ford tractor
sitting vacant in a field waiting for the farmer to return.

8.

She Beckons from Bondi

I stand with wind whipping
my hair on a boat
bound for Bondi.
Watching the water –
Colbalt waves dancing
with emerald reflections.
My first visit to the famous
"Sun-kissed" beach and I allow
the experience to absorb
into my pale skin.
I am mesmerized –
by the sea and the way it sings.
I appreciate the ancient
necessity for gods and goddesses –
Surely something pure
and holy must make the water
dance so mysteriously.
I close my eyes – the boat
rocks matching the heart beat
of the sea goddess. I can see
Her, and she knows me.

Acknowledgements

Three poems in this collection previously appeared in the literary journals *Amaranth* "She Beckons from Bondi," *West Texas Review* "Quakertown Trolley, Est. 1898," and *Common Ground Review* "New World."

Portions of the introduction appear in the essay collection *The Unfurling Frond*.

Thank you to Nick Courtright at Atmosphere Press for taking a chance on this unusual multimedia, hybrid-genre text. Thank you to Alexis Kale for the wonderful editorial advice – I don't know if you saw my ecstatic smile when you suggested I add more theory.

A sincere thank you to Brenda Miller for introducing me to the magic of photography and creative nonfiction. The genesis of this book was the final project for Brenda's seminar class at Western Washington University.

Thank you to Casey Corcoran, who so graciously read my twenty-page draft while other peer review partners only had to read five to seven pages. I appreciate your time and your insightful suggestions.

Thank you, Emily Allen, for working on the original book design. Your creativity and talent helped transform my Word document and concept for a hybrid text into a masterpiece to present to my class and Brenda. (And now the world.)

Mahalo nui to Kristiana Kahakauwila for answering my frantic texts and reminding me I've been on this journey of the spiral before.

A shout out to Matthew Taron for pointing me back to Didion.

Athena Roth, your continued encouragement and belief in me helped me to take this next step to present this work out into the world. Thank you, frond. (Yes, frond, that's not a typo.)

In loving memory and eternal appreciation to my Myrtle girl. Your support of my writing was constant. You never left my side through drafts, research, and revisions. In fact, you were by my side when I found the photo of Mary. You were a balm on my soul. I miss you.

Thank you to my husband Geoffrey; your unending support means the world to me. Thank you for joining me on this journey.

Colleen Clemens, thank you so much for listening and being my sounding board as I bounce around ideas. Thank you for taking a day to drive around the back roads of Pennsylvania with me. You took me to Mary, and for that, I'm forever grateful.

A special thank you to Aunt Barbara for sharing photos of Mary and Ruth with me.

This book would not exist without you, Mom and Aunt Pat. Thank you for seeing Ruth in me and being bold enough to say it. Thank you for the countless emails and text messages and for your patience with all my questions.

Rebecca Beardsall (MA, Lehigh University; MFA, Western Washington University) works at Western Washington University. Rebecca is the prose editor at *Psaltery & Lyre* and the nonfiction editor at *Minerva Rising Press*. She has more than twenty years' experience in freelance writing in the United States and abroad. Her poetry and essays have appeared in *Thimble, SWIMM, West Texas Review, Two Cities Review, The Schuylkill Valley Journal, Amaranth, Common Ground Review, Poetry NZ,* and *Rag Queen Periodical.* She wrote and co-edited three books, including *Philadelphia Reflections: Stories from the Delaware to the Schuylkill.* Find her at: rebeccabeardsall.com